The Young Scientist Investigates

Everyday Chemicals

Contents

Chemicals

Copper sulfate crystals

There are thousands of kinds of chemicals. A few are shown in the pictures.

Many chemicals look alike, but scientists know the difference. Scientists have given each chemical a name of its own. Some of these chemicals are solids, some are gases, some are liquids. One of the most important chemicals is water. This liquid chemical is dealt with in another book in this series. Some chemicals are solutions. A solution is a liquid with a solid, or a gas, or another liquid dissolved in it.

But where do all these chemicals come from? We get some chemicals from the sea. We get some chemicals from the earth. Some chemicals are made from plants. A few chemicals are made from the air. Many chemicals have been made by scientists in factories or laboratories.

Chemicals help us in many ways. Some chemicals help to make us better when we are ill. Medicines are chemicals. Some chemicals kill germs, while others stop food from going bad. Some chemicals make plants grow better. Some chemicals help us to make metals, plastics, and other materials. All the time scientists are making new chemicals that may help us in our everyday life.

The Young Scientist Investigates

Everyday Chemicals

by
Terry Jennings

CHILDRENS PRESS ®

CHICAGO

Library of Congress Cataloging-in-Publication Data

Jennings, Terry J.
    Everyday chemicals / by Terry Jennings.
        p.   cm. — (The Young scientist investigates.)
    Summary: An introduction to the many kinds of chemicals,
describing some of their uses to man. Includes study questions,
activities, and experiments.
    Includes index.
    ISBN 0-516-08401-1
    1. Chemicals—Juvenile literature. 2. Chemistry—
Experiments—Juvenile literature. [1. Chemicals.] I. Title.
II. Series: Jennings, Terry J. Young scientist investigates.
QD35.J53   1988
540—dc 19   88-22888
                                                                CIP
                                                                AC

North American edition published in 1989 by Regensteiner
Publishing Enterprises, Inc.

© Terry Jennings 1984
First published 1984 by Oxford University Press

Printed in the United States of America
1 2 3 4 5 6 7 8 9 10 R 98 97 96 95 94 93 92 91 90 89

Chemicals and heat

Ice cubes melting

Many substances change when they are heated. When ice cubes are warmed, they melt and form water. When a candle is lit, some of the wax melts and forms a liquid. Even iron will melt if it is heated to a very high temperature. But all of these things can be made solid again. If the water, liquid candle wax, and iron are cooled, they will turn back to the solid they came from.

Toasting bread

But some substances change forever when they are heated. It is not possible to get the original substance back again. When you toast a piece of bread, it turns brown or even black. The heat has changed the bread into a different substance. When we boil an egg, or bake a cake, we make something different. Cooked meat is very different from raw meat. Bread is very different from dough.

Making glass bottles

Scientists make many new chemicals by heating substances. When chalk is heated a new chemical is made. This chemical is called lime. If sand, soda, and lime are heated together, glass is made. Another way scientists make new substances is by mixing certain chemicals together.

sand
(silica)

soda
(sodium
oxide)

lime
(calcium
oxide)

Chemicals from coal

Coal is one of the most useful substances. Not only do we burn coal for fuel, we also obtain many chemicals from it. Dyes and paints are made from coal. So are aspirin, some fertilizers, and explosives. Some plastics, disinfectants, and glues are made from coal. More than 2,000 different chemicals can be made from coal.

The coal came originally from plants. Millions of years ago much of the land was covered by dense forest. The trees and other plants died and were covered by mud and sand. Very slowly the mud and sand were turned into rock. As more and more rock pressed down, the trees and other plants were slowly turned into coal. We can sometimes find the fossils of the plants that formed coal.

Huge machines are used to dig up the coal when it is near the surface. The coal deeper down is dug by miners working in tunnels. Much of the coal we mine is heated in large ovens. When coal is heated, a change takes place. Coke is formed. Coke is a better fuel than coal because it burns with less smoke. When coal is heated to form coke, a lot of gases are produced. These gases contain coal tar. From the coal tar and the gases, many of the useful chemicals mentioned above are made.

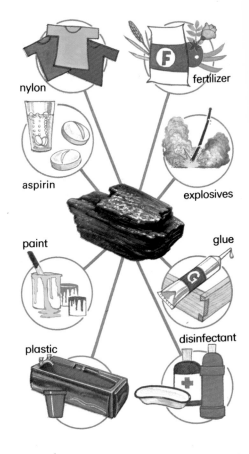

nylon

fertilizer

aspirin

explosives

paint

glue

plastic

disinfectant

Coal-cutting machine

Mining coal

Chemicals from oil

Oil tankers being loaded

Oil drilling rig

We get many chemicals from oil. Oil is an important fuel. It is used to drive trucks, buses, and some trains. Gasoline is made from oil. Gas is used to work the engines of cars. Kerosene from oil is a fuel for jet engines. Oil is also used to heat our houses, shops, offices, and factories.

Oil was formed from the bodies of tiny sea plants and animals that lived long ago. Rocks later pressed down on these tiny plants and animals, turning them into oil. To reach the oil, deep holes have to be drilled into the ground. Sometimes the drilling takes place under the sea. If the sea is deep, a large drilling rig is used to drill for the oil. Big pipes or ships carry the oil to the land.

Oil is not only an important fuel. Many other useful chemicals can be made from it. These chemicals help us make plastics and artificial rubber. They are also used to make candle wax, paints, lipstick, and perfumes. Many of the chemicals farmers use to kill weeds and insect pests come from oil. So do detergents.

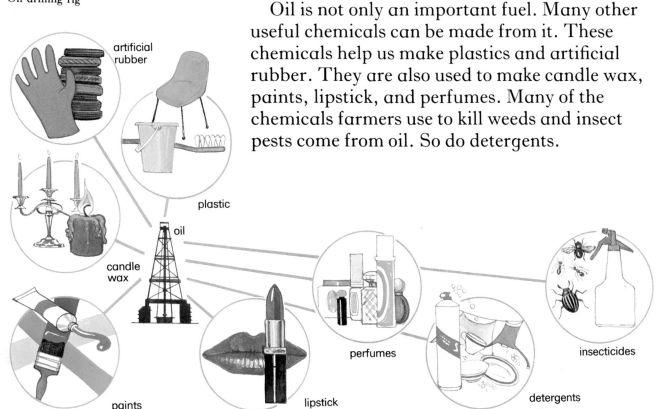

artificial rubber

plastic

oil

candle wax

paints

lipstick

perfumes

detergents

insecticides

Chemicals from the air

Although we cannot see it, air is a mixture of gases. One of these gases is oxygen. About one-fifth of the air is oxygen. All living things need oxygen.

Some people work in places where there is not much air. There is not much air near the top of very high mountains. So mountain climbers often take bottles of oxygen with them. Divers working under water also need air or oxygen to breathe. Some divers carry oxygen in bottles on their backs. Other divers get air, containing oxygen, from a tube. The tube goes to a pump on a boat.

In space there is no air at all. So astronauts take oxygen with them. Firemen sometimes have to carry bottles of air when they put out very smoky fires. There may be so much smoke that there is little oxygen for the firemen to breathe. Sometimes when people are very ill, the doctor gives them oxygen to breathe.

Fires need oxygen to burn. If something will burn in air, it will burn even better in pure oxygen. Oxygen mixed with other gases is used in special torches that cut or weld metals such as steel. Some oxygen also is used to make steel. Green plants make oxygen in sunlight when they make their food. If it were not for green plants, the oxygen in the air would soon be used up.

Mountain climber breathing oxygen

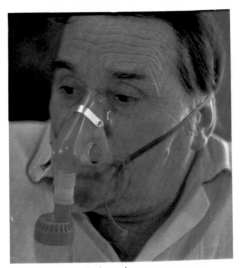
Very ill person being given oxygen

Cutting metal with the help of oxygen

Carbon dioxide

Another gas in the air is carbon dioxide. There is only a little carbon dioxide in the air, even though animals breathe out this gas.
When things like wood, paper, coal, oil, and gas burn, they make carbon dioxide.

Fire extinguisher being used

Some fire extinguishers also make carbon dioxide. Fires cannot burn when there is a lot of carbon dioxide around. The carbon dioxide forms an invisible blanket over the fire. The fire goes out because it cannot get enough oxygen to burn.

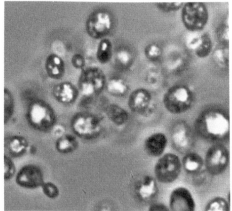
Yeast seen with a microscope

Bread and many cakes get their light texture because of carbon dioxide. Yeast, a tiny plant in the bread dough, makes carbon dioxide. The carbon dioxide makes all the little holes in the bread. A chemical called baking powder is put in some cakes. As the cakes are cooking, the baking powder gives off carbon dioxide. The carbon dioxide makes the cakes light and fluffy. Cola and other fizzy drinks contain a lot of carbon dioxide. These drinks are fizzy because the carbon dioxide in them tries to escape when the bottle cap is undone.

Holes in bread

Rows of lettuce in a field

Green plants use carbon dioxide to help make their food. They use carbon dioxide from the air and water, and mineral salts from the soil. Because green plants use carbon dioxide to make their food, there is never very much carbon dioxide in the air.

Acids and Bases

Two common acids: lemon juice and vinegar

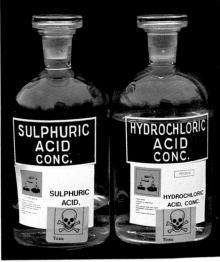
Bottles of acids

Two of the most important kinds of chemicals scientists use are acids and bases. Many acids are made by living things. The word acid means sour. Lemon juice is sour because of the acids in it. Vinegar contains a weak acid. It tastes sour.

Many acids are made by scientists from other substances. A lot of these acids are poisonous. They can also burn the skin or clothes. Some acids even dissolve metals. One very strong acid made by scientists is sulfuric acid. This acid is the one found in car batteries. It will burn the skin and clothes. If sulfuric acid gets on the metal of a car it will dissolve it. Sulfuric acid is used in making some fertilizers. It is also used in dyes, plastics, and other chemicals.

Another strong acid is hydrochloric acid. Scientists often use this to make new chemicals. Hydrochloric acid also is found in our stomachs. The acid helps us to digest our food. Too much acid in the stomach is bad for us. It causes indigestion.

Truck carrying phosphorus

Indigestion can hurt

Bases

There are many kinds of bases. Most of them are made by scientists. One very strong base is sodium hydroxide or lye. It is used to make soap and paper. Lye is a dangerous chemical. It should not be touched or tasted. Lye will burn the skin and make holes in clothes and metals.

One base that is safer to use is washing soda or sodium carbonate. We use sodium carbonate in the home. It helps soap to lather more easily and to dissolve grease.

Another base used in the home is baking soda. It is used in cooking. Baking powder also contains baking soda. When baking soda is heated it gives off carbon dioxide gas. The carbon dioxide makes cakes lighter. Some people take baking soda to help stop indigestion. It helps to dilute or weaken the acid in the stomach.

When chalk or limestone is heated in a kiln, it forms lime. Lime is a base. Farmers and gardeners put lime on the soil to make it less acidic. Many plants grow better in soil that is less acidic. Scientists make many new chemicals by mixing acids and bases together.

Using washing soda (sodium carbonate) to clean a greasy floor.

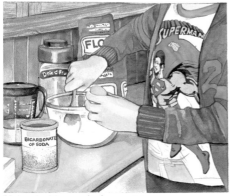
Using baking soda for cooking

Using lime on a garden

9

Do you remember?

(Look for the answers in the part of the book you have just been reading if you do not know them.)

1   What are chemicals like?

2   Where do chemicals come from?

3   What happens when candle wax is heated and then cooled again?

4   What are two examples of ways in which new substances can be made by heating other substances?

5   How was coal formed?

6   Name three of the chemicals that can be made by heating coal.

7   Why is coke a better fuel than coal?

8   How was oil formed?

9   Name three of the chemicals we can get from oil.

10   What is air made of?

11   Why must we have oxygen?

12   If something burns in air, how will it burn in pure oxygen?

13   How can carbon dioxide be made?

14   How do green plants use carbon dioxide?

15   How does carbon dioxide make bread and some cakes better?

16   Why is cola fizzy?

17   What does the word "acid" mean?

18   What is the acid called that is used in a car battery?

19   What happens when we have too much acid in our stomach?

20   Why do farmers and gardeners put lime on the soil?

Things to do

1   **Make a collage.** Make a collage by using cut-out pictures to make one large interesting picture about chemicals and the way they help us.

2   **Invisible ink.** Invisible inks are made from chemicals that, when you heat them, or treat them with other chemicals, change and become colored.

One simple kind of invisible ink is lemon juice. To see how it works, use a fine paintbrush to write a message with lemon juice on a sheet of white paper. Leave the paper in a warm place for an hour or two until the message is invisible.

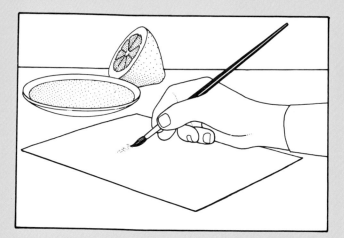

Vinegar and baking soda make carbon dioxide. The jar fills with carbon dioxide.

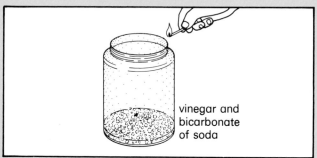

vinegar and bicarbonate of soda

To make the message appear, ask a grown-up to set an oven for you at 350°F. Ask the grown-up to leave the sheet of paper in the oven for ten minutes. The temperature is not high enough to burn the paper, but the heat changes the lemon juice to a brown substance.

3   Water and dissolved air. Take a clean jam jar and put some cold water in it. Let the jar of water warm up by placing it near a radiator or a sunny windowsill. Can you see tiny bubbles of air rising in the water? The air was dissolved in the water but comes out as you heat it. Fish use their gills to breathe the air that is dissolved in water.

4   Making carbon dioxide. To make carbon dioxide gas you need some baking soda, some vinegar, and a large jar such as a candy jar.

Put half a cup of baking soda into the large jar. Add half a cup of vinegar to the baking soda. The mixture bubbles and fizzes. The bubbles are carbon dioxide gas.

Strike a match. Carefully, hold the flame to the mouth of the jar. The match goes out because fire needs oxygen to burn. A match cannot burn in carbon dioxide.

Hold a lighted match to an empty jar. Does it go out?

5   Making unleavened bread. Unleavened bread is bread that is not made to rise by carbon dioxide from yeast. If you make some you can see what a difference carbon dioxide makes to the ordinary bread we eat.

To make some unleavened bread, put about ¼ cup of plain flour in a cup or small bowl. Add water, a little at a time, and mix it thoroughly to make a stiff paste or dough. Roll the dough into a ball on a floured working surface. Flatten the ball slightly and place it on a greased baking tin.

Ask a grown-up to help you bake your loaf in a hot oven at 450°F. for ten to fifteen minutes, or until it is golden brown.

When it is cool, taste your unleavened bread. This is similar to the bread people ate long ago, before they discovered how to make risen bread using yeast.

dough

stick

camp fire

baking unleavened bread

If you can make a campfire in the back-yard, you might like to bake your own bread over that. Make the dough into a sausage shape. Find a stick from a bush or a tree, and strip the bark from it. Wrap your dough around the end of the stick in a spiral. Cook the dough over the glowing embers of the fire, as seen in the picture. This rather burnt bread must be even more like the bread of olden times.

6   Growing yeast. As we saw on page 7, yeast is a tiny plant. It makes carbon dioxide that produces the little holes in bread. You can see the way yeast behaves if you half-fill a plastic bottle with warm water and add a package of yeast. Stir the mixture. Now add two teaspoons of sugar and stir the mixture.

Lightly push a cork in the bottle. Do *not* use a screw cap. Stand the bottle in a bowl of warm water. Look at the bottle from time to time over the next hour or so. What happens to the mixture in the bottle? What happens to the cork? Why is this? Smell the liquid inside the bottle. What do you notice? Do not drink the liquid, though.

If you have a microscope you could put a drop or two of the liquid from the bottle onto a slide. Focus carefully. Can you see the tiny yeast cells? If you watch them carefully you may be able to see them dividing as they grow and multiply.

7   Green plants make oxygen. In sunlight green plants make oxygen. It is very difficult to see the oxygen being made by a land plant, but you can see the oxygen coming from a water plant.

If you look in a pond or lake on a sunny day, you can often see bubbles of oxygen coming up to the surface from the leaves of water plants.

You can also see the oxygen if you get a small aquarium or large transparent glass jar. You need a glass or plastic funnel that is also transparent. The funnel should fit inside the jar or aquarium.

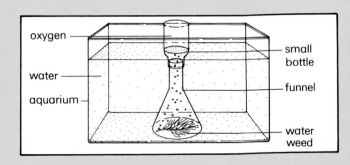

oxygen

water

aquarium

small bottle

funnel

water weed

Fill the jar or aquarium with water so that it is deep enough to completely cover the funnel. Before you put the funnel into the water, place a piece of water weed, such as pondweed into it. If you cannot get any water weed, a piece of mint works equally well.

Lower the funnel and plant into the water. Then fill a small transparent glass or plastic bottle or tube with water. Put your thumb over the top. Turn the bottle upside down with your thumb over the top, and slip the bottle onto the top of the funnel as shown in the picture. The bottle should stay filled with water.

Stand the jar or aquarium on a sunny windowsill. As the plant makes oxygen you will see the little bubbles rise and collect in the bottle or tube.

When the bottle is completely filled with oxygen, carefully lift the bottle out of the water. Still keep the bottle upside down. Ask a grown-up to light a match for you and hold it. When the match is burning well, blow out the flame and put the red-hot piece of wood quickly into the bottle. The oxgen will probably make the match burst into flames again.

Things that will burn in air will burn even better in oxygen.

8   A model fire extinguisher. Is there a fire extinguisher in your school? Do you know how it works?

You can make a model fire extinguisher. Take a clean bottle about 8 inches high. Find a cork that fits the bottle. Make a hole in the cork and put a short length of drinking straw through it. If you cannot find a suitable cork, you could make a plug from clay.

Put a teaspoon of baking soda in a paper napkin. Get a piece of thread about 10 inches long and tie up the paper napkin with the baking soda inside. Leave a piece of thread to hang the napkin.

Put three tablespoons of vinegar in the bottle. Keep the bottle upright and lower the paper napkin full of baking soda a little way into the bottle. Carefully put the cork on the bottle so that the cork (or clay) traps the thread and holds it in place.

Light a candle on a saucer, or in a candlestick (CAREFUL!).

Tip the bottle. The vinegar will react with the baking soda, making carbon dioxide. The carbon dioxide comes out of the bottle through the drinking straw.

Hold the end of the straw near the candle flame. The flame will go out.

A fire extinguisher contains two substances similar to this that react to make carbon dioxide.

Do *not* point your fire extinguisher at anyone. Keep your own eyes and face away from it.

Salt

Salt is a solid chemical. We use it when we cook. We also sprinkle salt on our food. There is salt in our sweat and tears, and salt in our blood. When we sweat, cry, or go to the bathroom we lose some of this salt. We have to eat salt in our food to make up for the salt we have lost.

There is salt in the soil and some in fresh water. There is a lot of salt dissolved in the sea. When it rains on land, the water soaks into the ground. Salt, which is in the soil, dissolves in the water. The water seeps from the ground into streams and rivers. The streams and rivers flow down to the sea. When seawater dries up, the salt is left behind.

In some hot countries salt is made by letting seawater dry in the sun. Some other countries get salt from salt mines. The salt in salt mines was left by seas that dried up long ago. There are salt mines in Great Britain, Germany, Poland, and the United States.

In Poland, Germany, and Britain, salt is mined in the same way as coal, but in Britain water is usually pumped down the salt mine. The water dissolves the salt, forming salt solution. The salt solution is pumped up to the surface. The water evaporates and salt is left.

Packets of salt

Baby crying

Getting salt from sea water

Inside a salt mine

Using salt

Salt being spread on an icy road

Bleach and sodium carbonate

Using bleach in the kitchen

Swimming pool containing chlorine

Only some of the salt that is mined is sprinkled on our food. A lot of salt is used on the roads in winter to melt ice and snow. Some salt is used to preserve food to stop it from spoiling. Bacon and some kinds of fish can be preserved by using salt.

Much salt is used to make other chemicals. One of the chemicals made from salt is sodium carbonate. Another is lye or sodium hydroxide. As we saw on page 9, lye is a very strong base. Lye is used to make soap and paper.

Chlorine is also made from salt. Chlorine is a green gas. It is used to make the bleach we use in our kitchens. Chlorine is put into swimming pools to kill germs. In most towns, chlorine is also put in drinking water to kill any germs. Many disinfectants and antiseptics contain chlorine.

Large quantities of salt are used in making pottery, plastics and dyes. Salt is also used in making leather from animal skins. Even butter and margarine are made with the help of salt.

Sugar

Sugar is a chemical that comes from plants. All green plants contain some sugar. But the sugar we eat comes from two plants – sugarcane and sugar beet. These two plants contain a lot of sugar.

Harvesting sugarcane

Sugarcane is a large grass with thick stems. The stems are about 2 inches thick. Sugarcane is grown in the West Indies, Australia, and other places with a hot, wet climate. To obtain the sugar from sugarcane, the stems are first cut into small pieces. These are then crushed so that the juice runs out. The dark gray juice is boiled. After a time the juice gets thicker. Soon lumps of sugar begin to form. The mixture is spun around in big drums. The thick sticky juice, called molasses, runs off. Molasses is used in making candy, rum, and alcohol. The sugar is left in the drum.

Sugar beet plants have large, fat roots. The roots contain a lot of sugar. Sugar beet is grown in some parts of Europe and the United States, where the weather is cooler. When the sugar beet is harvested in autumn, the roots are crushed in machines. The crushed roots form a pulp that is boiled. When the liquid dries, sugar crystals form. Sugar is used to make about 200 other chemicals. One of them is alcohol.

Sugar beet being harvested

Sugar being put into packets

Chalk

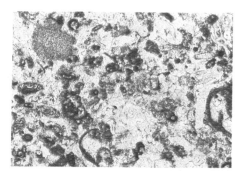
Chalk seen through a microscope

Chalky cliffs

A piece of gypsum

Chalk is a chemical that is dug out of the ground. Chalk is a soft, white rock. It is a kind of limestone. Chalk is partly made up of the shells of tiny animals. Millions of these tiny animals once lived in the seas. When the animals died, their shells piled up on the seabed. Layers of mud and sand formed over the tiny shells. Slowly the layers of shells were turned into chalk. Later the chalk was raised above sea level. We know when we see chalk in a quarry, on a railway bank, road cutting, or in a cliff, that that piece of land was once at the bottom of the sea.

Chalk is used to make many other substances. Some is heated in a kiln, to make lime. Lime is put on the farmer's fields to make the soil less acidic. Many fertilizers contain lime. Lime is also used to make glass. A lot of chalk is heated with clay in kilns to make cement.

Chalk also is used to make whiting. Whiting is used in paints, putty, and in the cleaning and polishing powders used in the home. Blackboard chalk is not really chalk at all. It comes from a soft rock called gypsum. Gypsum is also used to make wall plaster and plaster of paris.

Farmer spreading lime on fields

17

Dyes and colorings

Most of our clothes are colored using chemicals. These chemicals are known as dyes. A lot of the foods we buy in cans, bottles, or jars have had colors added to them. So have some sweets. These colorings are also chemicals.

Long ago people found out how to make colored substances from plants, small animals, and rocks. They used these colors to paint pictures on the walls of caves. Long ago people also made dyes to color their clothes. Most of the dyes were made from plants. A few dyes were made by boiling small animals such as shellfish from the seashore. Cloth used to be dyed in big tubs. Today big machines are used for dyeing cloth.

Wool being dyed

People used to make dyes from dog whelks like these

Most of the dyes and colorings we use nowadays come from oil or coal. The first artificial dye was made from coal. It was mauve. Plants, animals, and rocks could be used to make only a few colorings or dyes. Every kind of color can be made from coal or oil.

Soaps and detergents

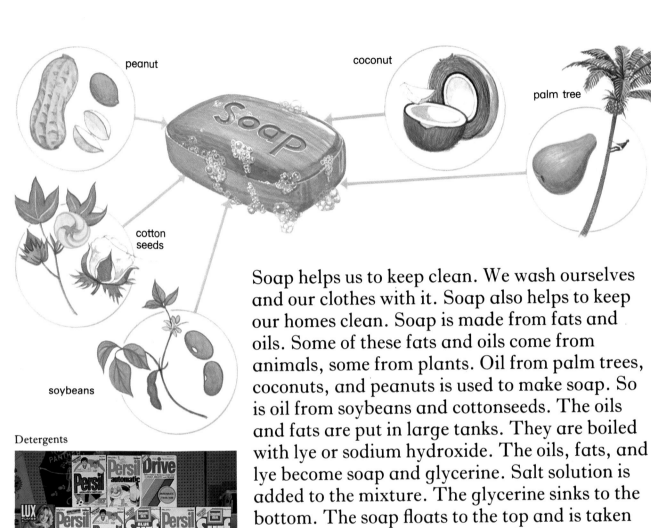

peanut

coconut

palm tree

cotton seeds

soybeans

Detergents

Soap helps us to keep clean. We wash ourselves and our clothes with it. Soap also helps to keep our homes clean. Soap is made from fats and oils. Some of these fats and oils come from animals, some from plants. Oil from palm trees, coconuts, and peanuts is used to make soap. So is oil from soybeans and cottonseeds. The oils and fats are put in large tanks. They are boiled with lye or sodium hydroxide. The oils, fats, and lye become soap and glycerine. Salt solution is added to the mixture. The glycerine sinks to the bottom. The soap floats to the top and is taken off. Sometimes perfumes and colorings are added to the soap. The soap is then cut and wrapped.

Detergents work in a similar way to soap. But detergents are made from chemicals that come from oil or coal. Some detergents are solids, some are liquids. Detergents are used for washing clothes and dishes. Many detergents contain chemicals to make clothes look whiter. The chemicals farmers use to kill weeds and insect pests also contain detergents.

Killing germs

Bacteria are very tiny plants. You cannot see them with your eyes. You can see them only with a powerful microscope. Bacteria are found in the air, in water, and in the soil. Other bacteria live in our bodies, on our clothes and skin, and on our food. Bacteria are everywhere. A few kinds of bacteria cause diseases. Many people call the bacteria that cause diseases, germs.

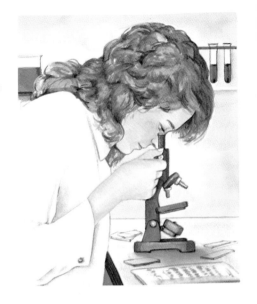

Scientists have made chemicals that will kill germs. We call some of these chemicals disinfectants. Most disinfectants are made from chemicals that come from coal. Many disinfectants are used in the house to kill germs in sinks, showers, bathrooms, and drains. The green gas chlorine is used to kill germs in drinking water and swimming pools. Chlorine is really a disinfectant.

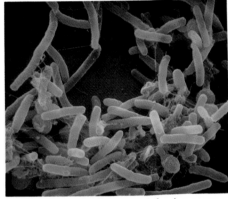

Bacteria seen under a powerful microscope

Antiseptics also kill germs. Antiseptics are put on wounds to kill germs. Doctors and dentists use antiseptics to kill germs on their instruments. Many antiseptics are made from chemicals that come from coal. Most antiseptics and disinfectants have to be used with great care. Ordinary soap and water will help kill germs. So will sunlight and fresh air. Our body can also make its own antiseptics. Tears, sweat, saliva, and blood all contain substances that help kill germs.

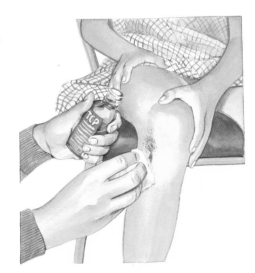

Farm and garden chemicals

Spreading manure

Farmers and gardeners use many chemicals. Some of these chemicals are used to make plants grow better. The chemicals are put on the soil and the plant roots take them up. Farmers and gardeners used to use a lot of manure. Manure is the waste from animals. Manure has a lot of plant foods in it. These plant foods are called mineral salts. But there is not enough manure for all the fields and gardens. So many farmers and gardeners use chemicals called fertilizers. Fertilizers are clean and they do not smell. They contain a lot of mineral salts that plants need.

Scientists have also made other chemicals to help the farmers' crops grow better. Weed killers are chemicals. They kill weeds but not the plants that the farmer or gardener wants. These weeds might otherwise take the water and mineral salts that the crop plants need. There are also chemicals that kill certain insect pests. These pests feed on the farmer's crops. The best of these chemicals kill the insect pests but do not harm useful insects like bees and ladybugs.

Using a helicopter to spray weed killer

Scientists also have made many medicines for the farmers' animals. These chemicals help cows, sheep, horses, pigs, and other farm animals grow and stay healthy.

Pollution

Chemicals can be very useful to us. If chemicals are used carefully, they are safe. Unfortunately some chemicals are doing harm. Chemicals given out in the exhaust fumes from cars and trucks are dirtying the air. We call this pollution. The air also is being polluted by chemicals in the smoke from factory chimneys. Even cigarette smoke is polluting the air.

Chemicals put on fields to make crops grow better have been washed into rivers and streams. They have killed the wildlife there. Chemicals put on fields to kill weeds and harmful insects have sometimes killed useful insects such as bees. Some beautiful birds, butterflies, and wild plants have also been killed. Some factories have dumped waste chemicals into rivers and lakes. This pollution has killed off the fish and wildlife. Chemicals also have been dumped into the sea. These chemicals might poison the fish we have to eat.

We shall have to be much more careful how we use chemicals in the future. We shall have to be much more careful how we get rid of chemicals when we have finished with them also.

Oil pollution from a shipwreck

A beach polluted with oil

22

Do you remember?

(Look for the answers in the part of the book you have just been reading if you do not know them.)

1 Where is there salt in our bodies?

2 How does salt get into the sea?

3 How is salt obtained from mines in Britain?

4 What are three ways in which we use salt?

5 Why is chlorine often put in swimming pools and drinking water?

6 Where does sugar come from?

7 Where are sugarcane and sugar beets grown?

8 Name one chemical that is made from sugar.

9 Where does chalk come from?

10 How was chalk formed?

11 What are two uses of chalk?

12 What are the chemicals called that we use to color our clothes?

13 What are the different kinds of coloring made from?

14 What is soap made from?

15 What are detergents made from?

16 What are bacteria?

17 What are disinfectants?

18 What do doctors use antiseptics for?

19 What are the plant foods called that are in manure and fertilizers?

20 Why do farmers and gardeners use weed killers?

21 What is pollution?

22 What has sometimes happened when chemicals put on fields have been washed into streams and rivers?

Things to do

1 **Mineral salts in the soil.** As we saw on page 7, there are mineral salts in the soil. Plant roots take up these mineral salts and the plant uses them to help make its food.

You can see these mineral salts if you put a handful of soil in a clean jar. Half-fill the jar with clean, cold tap water and stir the soil.

Take a circular filter paper or cut a circle from blotting paper. Fold the paper as shown in the picture on page 24.

Moisten the inside of the funnel with water and place the cone of paper inside it.

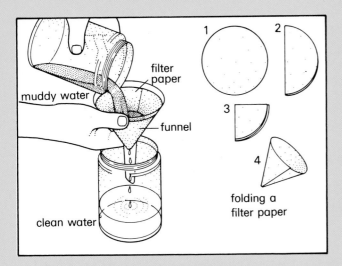

filter
paper

muddy water

funnel

clean water

1 2 3 4

folding a
filter paper

Add sugar to the water, a spoonful at a time, and stir. Keep adding sugar until no more will dissolve. There should be a little sugar left on the bottom of the jar.

Rest a pencil across the top of the jar. Hang a thin length of string from the pencil in a loop into the jar. Stand the jar in a warm place for a few days. What do you notice?

Can you get other substances to behave in the same way?

Stand the funnel over another clean jar and slowly and carefully pour the muddy water into the paper. The paper sieves the fine bits of soil from the water. We call this filtering. The water that comes out of the bottom of the funnel should be perfectly clear.

Put some of this clear water in a clean white saucer. Leave the saucer of water near a radiator or on a sunny windowsill. The water will evaporate and you will see the mineral salts left around the saucer. What color are the mineral salts?

You might be able to get two different soils and see which contains the most mineral salts. For your experiment to be fair, though, you must use the same amount of each soil and add the same amount of water to them both.

2   A necklace of sugar crystals. Take a clean jam jar and stand a metal spoon in it. Carefully pour hot water into the jar. The water should be just hot enough for you to put your fingers in it. If in doubt, ask a grown-up to help.

3   Growing crystals. With care and patience it is possible to grow large crystals. Sometimes crystals an inch long, or even longer, can be made. To begin with, try one of the chemicals that are sold by most chemists:

copper sulfate (makes blue crystals)
chrome alum (makes violet crystals)
potassium nitrate (makes white crystals)
magnesium sulfate or Epsom salts
(makes pearly white crystals)

Put ¾ of a pint of warm water into a clean jar or bowl. The water should have a temperature of about 122°F. (if in doubt ask a grown-up to help).

Add the chemical slowly, stirring all the time. Keep adding the chemical until no more will dissolve. You will then be able to see a little of the chemical at the bottom of the solution. Cover the jar or bowl with a cloth or sheet of paper so that dust does not get into it. Leave the solution for 24 hours.

When the crystal has grown as big as possible, carefully remove it from the jar or bowl. Cut off the thread or hair. Dry the crystal with a piece of filter paper. Store it in a matchbox containing cotton. Label the box with the name of the chemical that made the crystal.

Which chemical gives the biggest crystals? Are all crystals of the same chemical the same shape? Are crystals of different chemicals different shapes?

Note: After handling chemicals or crystals, always wash your hands thoroughly, especially before you touch food.

After 24 hours there will probably be even more of the chemical at the bottom. Carefully pour off the clear solution into a clean container.

Put a little of the solution of the chemical in a clean saucer. Pour it to a depth of about ⅓ of an inch. Cover the main jar or bowl of the solution again. Leave the saucer overnight or as long as it takes for small crystals ¹⁄₁₀ of an inch long to form in the bottom.

Choose the best shaped crystal you can find in the saucer. Lay the crystal on a clean piece of filter paper. Carefully tie a piece of thin nylon thread or a long hair around the crystal. Tie the other end of the thread or hair to a pencil. Hang the small crystal into the jar or bowl of the solution. Stand the whole thing in a cupboard and keep the top of the jar or bowl covered. With luck, the crystal will grow bigger. Where does the crystal get the material with which to grow?

4 Sugarcane. Much sugarcane is grown in the West Indies. Find a map of the West Indies in your atlas. Copy the map into your book. Now find the name of each island and print it on your map. How are ships carrying sugar likely to get from the West Indies to the United States? Look at possible routes in your atlas. Which is the shortest route?

5 Chemicals and water. Collect some clean jam jars and fill them halfway with water. Into one jar put a level teaspoonful of sugar and stir it. The sugar spreads throughout the water and disappears. We say the sugar has dissolved and made a solution.

If the same amount of sand is put in a jar half filled with water, the sand sinks to the bottom. The water does not dissolve the sand, even if we stir it.

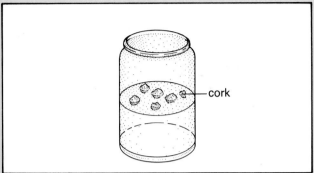

cork

If we put small pieces of cork in a jar of water, they float. Even if they are pushed under, they bob up. The water does not alter the cork.

If we put a little piece of clay in water and stir it, the water goes cloudy. This is because the clay breaks up into tiny pieces that remain suspended in the water. After a long time they may settle.

Now test some other substances to see whether they dissolve, sink, float, or remain suspended in the water. Does anything else happen? Some of the substances you might try are salt, stones, rice, rubber, washing detergent, syrup, raisins, instant coffee, cocoa, vinegar, mustard, pepper, and a small piece of soap.

Do *not* try any other substances without asking a grown-up. Do *not* taste any of the substances.

6 Watching color patterns. Take a filter paper.

Mark a big spot of color in the center of the paper with watercolor paints or felt-tipped pens. Draw some wide-colored bands (in different colors) around the center spot. Stand the paper over the open top of a jam jar. With a drinking straw or an eyedropper, drop water onto the center spot. Watch the pattern grow as water soaks through the paper away from the center spot.

Try making other patterns and designs of your own.

Dry your papers and make a display of them on your classroom wall.

7 Making dyes. Ask a grown-up to help you with this work.

For your dye making you will need two small saucepans and the use of a cooker or hot plate. You will also need some pieces of white cloth (each about four inches × four inches). If possible, get some pieces of white cloth made from cotton, wool, linen, rayon, and nylon.

An easy dye to start with is tea. Put 5 or 6 teaspoons of tea (or 5 or 6 teabags) in about a half pint of water in one of the saucepans. Boil this for at least five minutes. Stir the tea leaves or tea bags from time to time (CAREFUL!).

Turn off the heat, and when the saucepan is cool, strain the colored liquid into the other saucepan. Place one or two pieces of your cloth in this liquid (your dye) and put it back on the hot plate or cooker to simmer for 10 minutes.

Turn off the heat and carefully lift out your piece of cloth using a spoon. Rinse the pieces of cloth in cold water. Leave the cloth where it will dry quickly.

Do all kinds of cloth dye the same color?

Does your dye wash out in soapy water?

Does your dye fade in the sunshine?

Now you know how to make one dye, try to use other plant materials to get dyes. Some you might try include blackberries, red cabbage, beetroot, onion skins, peas, dandelion roots, cherries, coffee, tomatoes, or lichens.

Remember: some berries and leaves are **poisonous**. If in doubt, ask your teacher for help.

Make a display of different kinds of dyed cloth using different plant materials.

Experiments to try

IMPORTANT: The experiments described here are quite safe and most of the chemicals are harmless. However, never invent your own experiments or play with chemicals. Keep all chemicals away from younger children.

Always wash your hands thoroughly after handling chemicals, and especially before touching food.

Do your experiments carefully. Write or draw what you have done and what happens. Say what you have learned. Compare your findings with those of your friends.

1 Testing some common powders

What you need: Collect small quantities (about a heaping teaspoonful) of as many common powders as you can find. Some you might collect are sugar, salt, flour, pepper, baking powder, baking soda, cocoa, coffee, mustard, curry powder, starch, dried milk, gravy powder, soap powder, talcum powder, and scouring powder. If in doubt about using a substance, ask a grown-up.

You will also need some vinegar; a drinking straw or an eyedropper; water; some small aluminum cake dishes; a candle; a hand lens or magnifying glass; some small bottles; a pair of oven gloves; two bricks; a lid of a cookie tin.

What you do: Make a list of the powders you have collected. Then ask a friend to give each powder a number. Get your friend to put the samples in small bottles or on pieces of paper with only the number written on each. Ask your friend to put away the list until you have finished the experiment.

Look at one of the powders with a hand lens or magnifying glass. What color is it? What does it feel like? What does the powder smell like? *Do not taste any of the powders, though.*

Put a small amount of the same powder in a small bottle (such as an aspirin bottle). Add a little water to the powder. What happens? Does the powder dissolve? If so, what color is the solution?

Put a little of the powder in an aluminum foil dish and use the drinking straw or eyedropper to add a few drops of vinegar to the powder. What happens?

Lay the lid of a large cookie tin on the table and place two bricks on it as shown in the picture. Place a short length of candle on a small tin lid and put it between the two bricks. Light the candle and *carefully* use the oven gloves to rest an aluminum foil dish containing a little of the powder, over the candle. Rest the dish on the bricks and allow the powder to heat for a few seconds. Then blow out the candle. Look at the powder. How has the heat changed it?

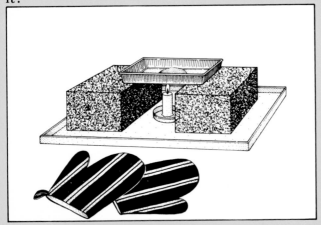

Make a table for your results like this and fill it in:

Powder number	Color and appearance	Feel	Smell	What I think the powder is

Carry out these tests for each of the powders in turn.

When you have finished the tests and filled in the table, ask your friend to let you see the list. How many of the powders were you able to identify correctly.

2 Air burning

What you need: A short piece of candle; a tupperware sandwich box; a small piece of clay; a tall narrow glass jar; a rubber band; a ruler.

What you do: Press a small piece of clay down in the center of the sandwich box. Then push the piece of candle into the clay so that the candle stands up. Pour water into the sandwich box so that the water comes half of an inch up from the bottom of the candle.

Carefully light the candle and lower the jar over the lighted candle. What happens to the candle?

Why do you think the water rises up inside the jar. What has gone from the jar that makes the water rise up?

Put the rubber band around the jar to mark the water level.

Now take the jar away and relight the candle. Lower the jar over the candle. Does the water rise to exactly the same level as before?

Measure how far the water has risen. Measure the total height of the jar above the level of the water in the sandwich box. Roughly what percentage of the jar has the water filled?

3　Cleaning dirty coins

Some chemicals can be used to clean metals. In this experiment we shall try to clean some dirty pennies.

What you need: 2 jam jars, plastic beakers or yogurt containers; vinegar; salt; a plastic spoon; some dirty pennies.

What you do: Pour some vinegar into one of the containers. The vinegar should be about ⅓ of an inch deep in the container. Add 1 level teaspoon of salt to the vinegar. Use the spoon to stir the salt and vinegar together.

Put some cold water into the other container.

Put one of the dirty coins in the salt and vinegar. Leave the coin there for a minute or two. Now use the spoon to take the coin out of the salt and vinegar. Wash the coin in the container of water.

What color is the coin now?

Will salt and water clean pennies?

Will vinegar on its own clean pennies?

Will the salt and vinegar clean nickels?

4　Separating salt and sand

Often scientists have to separate different chemicals from mixtures. Ask your teacher to mix equal parts of clean sand and salt for you, and to give you a little of the mixture. Can you separate the two substances?

What you need: Sand-salt mixture; 2 clean jam jars; funnel; filter paper; clean saucers; warm water (about 122°F); spoon.

What you do: Put a little of the mixture in a clean jam jar. Add a little warm water to the mixture and stir it thoroughly. What happens?

Now take the filter paper. Fold the filter paper as shown in the picture on page 24 and fit it inside the funnel. You will find the paper stays open if you dampen the inside of the funnel. Stand the funnel on a clean jam jar.

Slowly and carefully pour the sand-salt mixture and water into the paper inside the funnel. What happens? What is left in the funnel? What goes into the jar underneath the funnel?

Now pour some of the liquid from the jar under the funnel into clean saucers. Stand the saucers in a warm place and leave them there for a few days. What happens? What is left?

Why is it possible to separate sand and salt in this way? Is it possible to separate all kinds of mixtures of substances in this way?

5　How can we separate different colors?

What you need: Some white filter paper; a clean jam jar; a pencil; different colored inks, felt-tipped pens, dyes, paints, and fruit and vegetable juices.

What you do: Cut a strip of white filter paper about 8 inches long and 1 inch wide. Draw a line across the strip about an inch from one end. Use black ink or a black felt-tipped pen to draw the line.

Put a little water in the jar and lay a pencil over the top of the jar.

Hang the strip of paper over the pencil so that the end of the strip where the ink line is dips a little way into the water as shown in the picture.

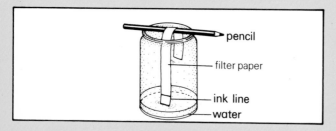

Watch the water soak up the paper. What happens to the ink line? How many different colors go to make up the black ink? What are they?

Try this experiment with other colored inks, felt-tipped pens, dyes, and paints. Use only inks, dyes, and paints that are *water soluble.*. If in doubt, ask a grown-up.

You might also try this experiment with tea, coffee, orange juice, blackberry and cranberry juices, the liquid from boiled cabbage, and the colorings used in cooking.

Dry each piece of filter paper used in the experiments and mount them in a book or on an index card. Write the name of the liquid tested against each piece of filter paper.

If you mix two different inks, dyes, or other colors together, can you separate them again using a strip of filter paper and the method used in this experiment?

6   Indicators

Scientists use special dyes called *indicators* to tell whether a substance is an acid or a base. The dye changes color according to whether the substance being tested is an acid or a base. If a substance does not change the color of the indicator, it is said to be neutral.

What you need: A clean raw beetroot; a kitchen knife; saucepan; water, clean jam jars; vinegar; sodium carbonate; baking soda; lemon juice; orange juice; milk; fizz drink; salt; indigestion tablets; Milk of Magnesia; tea; sour milk.

What you do: Cut the beetroot up into small pieces. Put the beetroot in the saucepan and cover it with about twice its own volume of water (if the beetroot makes a layer about an inch deep in the saucepan, add another 1½ inches of water).

Ask a grown-up to boil the beetroot in the water for about 20 minutes. Boil the beetroot until the water is colored deep red or purple.

When the liquid is cool, carefully pour it through filter paper in a funnel (see page 24). Collect the clear liquid in a jar. There should be enough for a lot of experiments, or enough for your friends to have some as well. If you do not want to use all the liquid in one day, store it in a dark bottle or a dark cupboard.

Put some of the indicator you have made in two clean jam jars. Fill the jars with the indicator to a depth of $\frac{1}{3}$ inch. Add a pinch of baking soda to the indicator in one jar and about a teaspoon of vinegar to the other. See what color the indicator turns. Remember baking soda is a base and vinegar is an acid!

Now test the other substances to see whether they are acids or bases. If the indicator turns the same color as it did with the vinegar, then the substance is an acid. If the indicator turns to the color it did with baking soda, it is a base. If the indicator does not change, the substance is neutral.

Try to make other indicators. Substances you might try to make indicators from include elderberries, blackberries, blueberries, red cabbage, and the petals of roses and other flowers.

7 Copperplating

Electricity can be used to make a new chemical from something else. For this experiment you need to buy some copper sulfate (a blue chemical). You can also use copper sulfate to make crystals (see page 24). But be careful as it is **poisonous**, and wash your hands thoroughly when you have finished the experiment.

What you need: A 4.5 volt battery; 2 pieces of thin insulated copper wire about 12 inches long; scissors; 2 paper clips; a piece of emery board; copper sulfate; a jam jar; water; a plastic spoon; 2 wooden pegs; a large nail or an old key.

What you do: Use the scissors to bare one inch of each end of both pieces of wire.

Fix a paper clip to one end of each piece of wire.

Rub the nail or key with an emery board, then join one of the wires to it. Look on the battery for the negative terminal (marked $-$). Use the paper clip to join the wire with the key or nail it to the negative terminal.

Join the other wire to the positive terminal on the battery (marked $+$). Fill $\frac{1}{3}$ of the jam jar with water. Stir in two teaspoons of copper sulfate. Throw away the plastic spoon after you have made the copper sulfate dissolve.

Dip the key or nail and the other wire into the copper sulfate solution. Leave it there for about 20 minutes. Do not let the key or the nail touch the other wire. You could keep them apart with two wooden pegs.

At the end of 20 minutes, look at the key or nail. What has happened to the blue copper sulfate solution? What has happened to the ends of the two pieces of wire?

Try to copperplate some other materials. Make lists of those you can and cannot copperplate.

Glossary

Here are the meanings of some words that you might have met for the first time in this book.

Acid: the word acid means sour. Some common acids include lemon juice, vinegar, and the sulfuric acid found in car batteries.

Bases: a group of substances that are the opposites of acids and can make acids less strong.

Antiseptic: a chemical used to clean wounds and to kill the germs on doctors' instruments.

Bacteria: tiny plants too small to be seen without a powerful microscope. Many bacteria cause materials to decay: a few cause disease.

Carbon dioxide: one of the gases in the air. Carbon dioxide is used by green plants to make their food. Fizzy drinks contain carbon dioxide, and yeast and baking powder make cakes lighter with this gas.

Chalk: a soft, white rock that is partly made up of the shells of tiny sea animals. The chalk used to write on blackboards is not really chalk at all. It comes from a soft rock called gypsum.

Chemicals: the thousands of substances, solids, liquids, and gases, that make up all living things and the world around them.

Chlorine: a yellowish-green gas often added to tap water at the waterworks to kill any germs. Chlorine is also put in the water of swimming pools to kill germs. Household bleach contains chlorine.

Detergent: a solid or liquid substance that is used in a similar way as soap to dissolve grease or to clean clothes. Detergents are made from chemicals that come from oil.

Disinfectants: chemicals used to kill germs in rooms and on clothes.

Dye: a chemical substance that is used to color wool, cloth, and other materials.

Evaporate: when water is heated it disappears into the air as water vapor. We say the water has evaporated.

Fertilizers: chemical substances put on the soil to make plants grow better.

Fuel: anything that will burn and produce energy.

Indicator: a special dye that changes color according to whether a substance is an acid or a base.

Lime: a base made by heating chalk or limestone in a kiln. Lime is often put on the soil to make it less acidic.

Mineral salts: the chemical substances that plants obtain from the soil and from manures and fertilizers, and that they use as food.

Neutral: a substance that is neither an acid or a base is said to be neutral.

Oxygen: the gas in the air that all plants and animals need. Oxygen goes into our blood when we breathe in. Oxygen helps our food to give us energy. It is also needed when something burns.

Plastics: a group of substances that can be shaped or molded into any form. Plastics are made from chemicals that come from oil.

Pollution: the dirtying or poisoning of air, water, or the soil, or the making of so much noise that it annoys other people.

Saliva: the liquid that is added to our food as we chew it.

Yeast: a tiny plant used in the making of bread, wine, and beer.

Acknowledgments

The publishers would like to thank the following for permission to reproduce transparencies:

Agricultural Lime Producers' Council: p. 17 (bottom); B.P. Oil Ltd: p. 5 (top, center), British Gypsum: p.17 (2nd from bottom); British Sugar plc: p. 16 (2nd from bottom, bottom); Cadbury Ltd: p.16 (2nd from top); John Cleare/Mountain Camera: p.6 (top); Bruce Coleman/WWF Mark Boulton: p. 22 (top); Bruce Coleman/Gene Cox: p. 7 (2nd from top); Bruce Coleman/Charles Henneghien: p. 14 (2nd from bottom); Bruce Coleman/Colin Molyneux: p. 6 (bottom); Crown Copyright: p. 8 (center); Color Library International: p. 18 (center); Compix: p. 16 (top); Nick Fogden: p. 2 (bottom), p. 7 (top); Geoscience Features: p. 2 (top right); Sally and Richard Greenhill: p. 6 (center); Imperial Chemical Industries (Mond Division) plc: p. 14 (bottom), p. 15 (top); Imperial Chemical Industries (Plant Protection Division) plc: p. 21 (2nd from bottom); Institute of Geological Sciences: p. 17 (top); Terry Jennings: p. 3 (2nd from top), p. 7 (2nd from bottom), p. 17 (2nd from top), p. 22 (2nd from top, 2nd from bottom, bottom); Lever Brothers Ltd: p. 19 (bottom); Massey Ferguson: p. 21 (2nd from top); National Coal Board: p. 4 (bottom); National Vegetable Research Station: p. 7 (bottom); Manfred Kage/Oxford Scientific Films: p. 20 (center); Picturepoint Ltd: p. 15 (bottom), p. 18 (top), p. 21 (bottom); Rosemarie Pitts: p. 9 (top, bottom), p. 21 (top); Science Photo Library: p.2 (top left); United Glass Containers: p. 3 (2nd from bottom); Zefa: p. 14 (2nd from top).